"The Greatest Inventors of All-Time"

Chapter 1

In the year of the 1800's there were famous African American inventors who created inventions in the United States of America, that are currently being used everywhere across the world. Some of these greatest inventions are common in the world today that people use in everyday life. For example: I know that these Inventors have created Inventions for people that live all over the globe. The following Inventions that are out here today were created by these great African American inventors and by other great American inventors as well and many other great inventors from all over the world.

The African Americans Inventors that I will be talking about are people that had created inventions in a well-known world. If it wasn't for these great inventors there wouldn't be inventions out here today for anyone to use across the country and world. African Americans had invented these inventions so that people would be able to do things that they would want to do. Going back into time during the early mid 1800's within Centuries Ago African Americans and other ethnicities had to sell these inventions, in order to be freed from slavery and had to give up their patent. Knowing about some of these great African American inventors that made these inventions, they will always be known for the work that they have done for people to use them. Back in the 1800's racism had existed, and during this time African Americans did not have the say so as to what they can do with their inventions, so the slave masters took the rights to their invention or the government had rights to their inventions. It was a little harder for the African Americans that were slaves besides the African Americans who were freed from slavery.

The following pages represent a unique and rare collection of African-American inventors and their inventions. This list originates almost entirely (with some exceptions) from the rare list compiled by Mr. Henry Baker in the late 1800s to early 1900s. According to research Mr. Baker was an African-American who worked for the U.S. Patent Office during these early years.

During this time African-Americans had little to no rights at all which made entering into contractual and legal agreements very hard but if not impossible. Since there were enough laws that prevented early African-Americans from participating in most legal processes those of contest against the racist white people had been concerned about ownership, claims, patent infringement, etc. Many African-American inventors lost or never gained legal rights to their inventions and for the ones that weren't as fortunate enough to obtain patents for their invention for example, were seldom recognized. There were slaves who became inventors automatically but lost the inventions to the slave master who owned them. During the slave trade there were many slaves that came from former Songhay Empire who were highly educated and was credited with teaching Caribbean and American farmers successful agriculture techniques. Slaves had invented various tools and equipment to lessen the burden of their day. Slave owners took their rights away so that slaves could not use them. Most slave inventors were nameless since they were owned by the confederate President Jefferson Davis. Following the Civil War, the growth of the economy was climbing because of the impact with inventions that slaves have invented. In 1913 there were at least 1,000 plus inventions that were patented by African Americans at this time.

Henry Baker took it upon himself to make sure that African-Americans who were awarded patents by the United States Government would be "unofficially " documented. This was unique and an important decision since the U. S. Patent office has never recorded culture or racial identity on patent applications. To accomplish this Henry Baker made a mark (only recognizable by him) on the forms of known African-American inventors who had submitted patents to the Patent Office. Using these marked forms he compiled one of the most important records in African-American history. These documents have become known as "The Henry Baker Papers." Without them we would have never known information about the magnitude with African-American contribution to society and the world. There were plenty of great ideas and inventions that was invented from lots other ethnicities as well. This shows and proves anybody can create and invent anything as long a person can put their mind to it.

Chapter 2

Henry T. Sampson

Born on April 22, 1934 and Died on June 4, 2015
The Inventor of Cellular Phone and Gamma Electric Cell   July 6,1971

George Washington Carver
Born in the year of 1860 and Died on January 5, 1943

George Washington Carver was born in 1860 in Diamond Grove, Missouri and despite early difficulties that have become hard in his life, he was one of the most celebrated and respected scientists in United States history. His important discoveries and methods enabled farmers throughout the south and mid-west and become profitable and prosperous.

George was born the sickly child of two slaves and would remain frail for most of his childhood. One night a band of raiders attacked his family and stole George and his mother. Days later, George was found alive and unharmed by neighbors and then traded back to his owners in exchange for a racehorse. Because of his frailty, George was not healthy enough to work in the field but he did possess a great interest in plants and was very eager to learn more about them.

Carver soon instructed nearby farmers about his methods by improving the soil and taught them how to rotate their crops to promote a better quality of soil. Most of the staple crops of the south (tobacco and cotton) had nutrients that were stolen from the soil, but these nutrients could be returned by planting legumes. Thus, in order to improve the soil, Carver instructed the farmers to plant peanuts, which could be harvested and fed to livestock. The farmers were ecstatic with the tremendous quality of cotton and tobacco that ended up growing but because of the amounts of peanuts that were harvested it was too plentiful and began to rot in overflowing warehouses. Within a week, Carver had experimented with dozens peanut, and this included milk and cheese. In later years he would produce more than 300 products that could be developed from the lowly peanut, including ink, facial cream, shampoo and soap.

Chapter 3

Lewis Latimer

Lewis was born September 4,1848-December 11, 1928 Invented and Improved the Efficient Light Bulb that are used in homes, buildings, and across the world. Today we use the same philosophy in electricity (Chamberlain, 2019) Lewis Latimer is considered to be one of the top 10 most important Black inventors of all time, not only for the sheer number of inventions he created and also the patents that are secured but also for the magnitude of importance for his most famous discovery. Latimer was born on September 4, 1848 in Chelsea, Massachusetts. His parents were George and Rebecca Latimer, both were runaway slaves who migrated to Massachusetts in 1842 from Virginia. George Latimer was captured by his slave owner, who was determined to take him back to Virginia. His situation gained great notoriety, even reaching the Massachusetts Supreme Court. Eventually George was purchased by abolition supporters who set him free.

Jan Matzeliger

Born on September 15,1852 and Died on August 24,1889 Invented an Automated Shoe Lasting Device

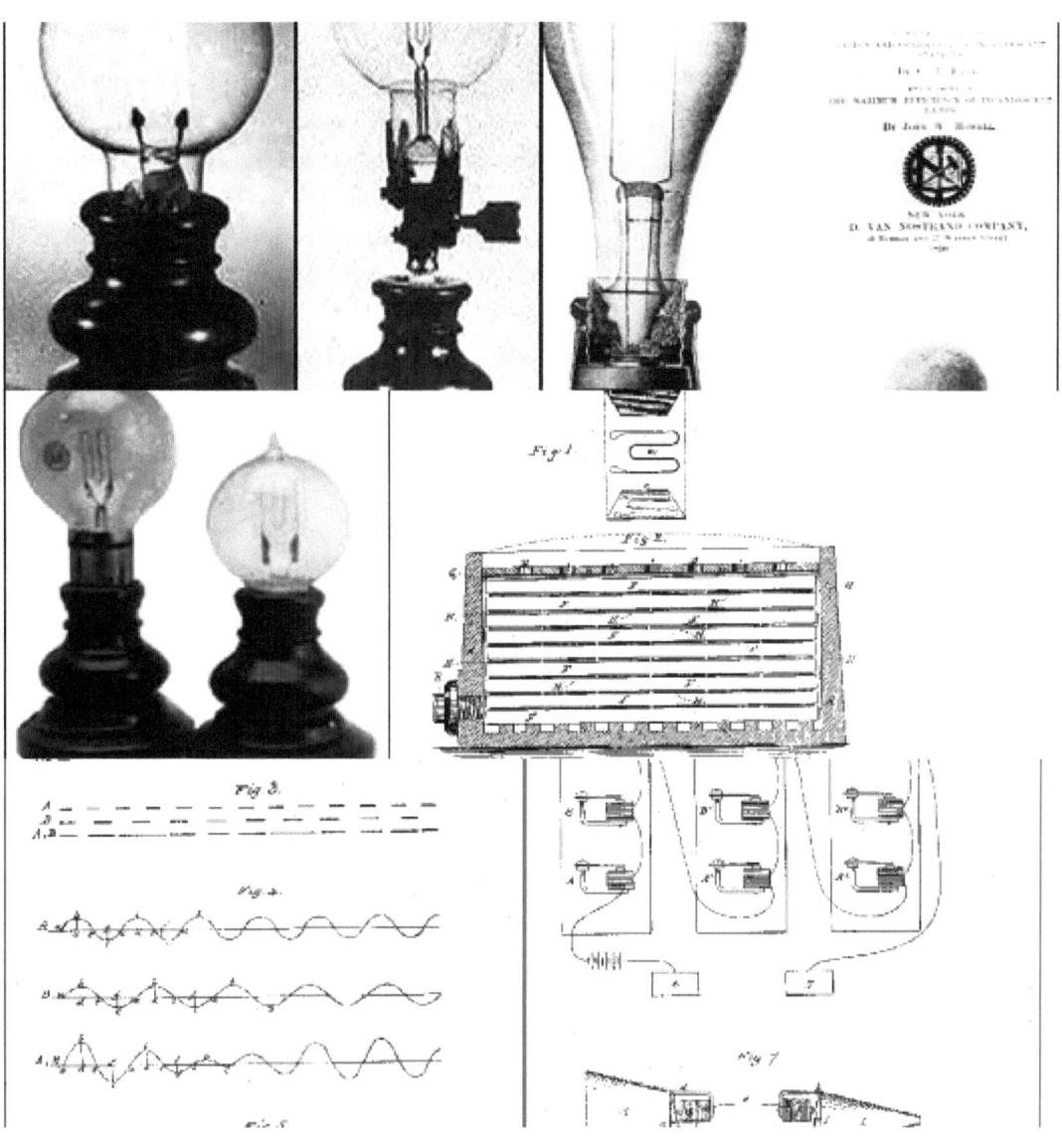

Chapter 4

Alexander Graham Bell (March 3, 1847 – August 2, 1922) was a Scottish-born scientist, inventor, engineer and innovator who is credited with inventing the first practical telephone. Bell's father, grandfather, and brother had all been associated with work on elocution and speech, and both his mother and wife were deaf, profoundly influencing Bell's life's work. According to (Chamberlain, 2019) His research on hearing and speech further led him to experiment with hearing devices which eventually culminated in Bell being awarded the first U.S. patent for the telephone in 1876. Bell is known to be considered the most famous with his invention an intrusion on his real work as a scientist and refused to have a telephone in his study. There were many other inventions who marked Bell's later life, that included groundbreaking work in optical telecommunications, hydrofoils and aeronautics. In 1888, Bell became one of the founding members of the National Geographic Society.

Madam C.J. Walker

Born on December 23, 1867 and Died on May 25,1919 Created a Black Hair Products Empire

Julian Pioneer

Born on April 11,1899 and Died on April 19, 1972
was a Pioneer in the Field of Synthetic Chemistry

Albert Richardson

Born on May 19, 1880 and Died on February 3, 1964

Inventor of the Butter Churn and a casket lowering device.
Albert Richardson was one a rare inventor who not only created numerous devices but created devices that were different to one another. Albert built the casket lowering device in 1894.

Until 1891 anyone wanting to make butter would have to do so by hand in a bowl. On February 17, 1891 Richardson patented the butter churn. The device consisted of a large wooden cylinder container with a plunger-like handle which moved up and down. In doing so, the movement caused oily parts of cream or milk to become separated from more watery parts. This was an easy way to make butter and forever change the food industry.

According to (Chamberlain, 2019) In 1894, Richardson saw a problem with the way the bodies of dead people were buried. It was common at that time to simply bury bodies in small, shallow graves or to try to lower their caskets with ropes into a deeper hole. Unfortunately, this required several people to work in unison to ensure that the casket was lowered evenly. Failure to follow directions could cause the casket to slip out of one of the ropes and to be damaged from hitting the ground. This invention was very significant at that time and is used in all cemeteries today.

Chapter 5

Andre Reboucas Invented an Early-version Torpedo. **Andre Reboucas was born in 1838** in Rio de Janeiro, Brazil. He was trained at the Military School of Rio de Janeiro and became an engineer after studying in Europe. After returning to Brazil, Reboucas was named a lieutenant in the engineering corps in the 1864 Paraguayan War. During the war, as naval vessels became more and more integral, Reboucas designed an immersible device which could be projected underwater, causing an explosion with any ship it hit. The device became known as the torpedo.

After his military career, Reboucas began teaching at the Polytechnical School in Rio de Janeiro and became very wealthy. According to (Chamberlain, 2019) He used his wealth to aid in the Brazilian abolition movement, trying to end slavery in Brazil. After growing disgusted with conditions in Brazil, Reboucas moved to Funchal, Madeira, off of the coast of Africa where he died in 1898.

Born 1849-May 10,1921 **Andrew Jackson Beard** hailed from Eastlake, Alabama, which was a small town outside of Birmingham. With the

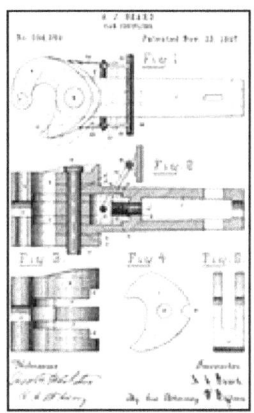 emergence of the railroad industry and its rapid expansion throughout the country, there were an alarming number of Railroad Men who suffered serious injuries to their arms and legs when they were crushed during manual style coupling of railroad cars. During manual coupling, a worker would have to attempt to precisely know the exact time in the moment when two railroad cars being pushed together that would be close enough for the worker to drop a metal pin between their connectors, and engaging the cars. If the worker was off by one second he might have been severely hurt with damage to his arm or leg – but many could have in fact had to undergo amputation if the injury were bad to their arms and legs. On November 27, 1897 Beard received a patent for a device he called the Jenny Coupler. According to (Chamberlain, 2019) The Jenny Coupler automatically joined cars by simply allowing them to bump into each other, or as Beard described it the "horizontal jaws engage each other to connect the cars." Beard ended up selling the rights to his invention for thousands of dollars and the railroad industry was revolutionized. During his lifetime, Beard received a number of other patents, including a steam driven rotary engine, and a double plow.

Benjamin Banneker was born in 1731 just outside of Baltimore, Maryland, the son of a slave. His grandfather had been a member of a royal family in Africa and was wise in agricultural endeavors. His father, Robert, was an African slave who purchased his freedom and his mother, Mary. Over the course of the next couple days, Benjamin repeatedly took the watch apart and then put it back together. After a few years of designing the clock and carving each piece by hand, including the gears, Banneker had successfully created the first clock ever built in the United States.

When a family friend died and left him a book on astronomy, a telescope and other scientific inventions. According to (Chamberlain, 2019) Banneker became great with the stars and the skies. In 1792, he developed his first almanac. Banneker sent a copy of his book to Thomas Jefferson, at that time the Secretary of State and in a twelve page later expressed to Jefferson that Blacks in the United States possessed equal intellectual capacity and mental capabilities as those Whites who were described in the Declaration of Independence. As such, he stated, Blacks should also be afforded the same rights and opportunities afforded to whites. This began a long correspondence between the two men that would extend over several years. He eventually moved back to France.

Banneker surprised them when he asserted that he could reproduce the plans from memory and in couple days did exactly as he had promised. The plans he drew were the basis for the layout of streets, buildings and monuments that exist to this day in Washington D.C.

According to (Chamberlain, 2019) Benjamin Banneker died quietly on October 25, 1806, lying in a field looking at the stars through his telescope. Nations around the world mourned his passing, viewing him as a genius and the United States' first great Black Inventor. In 1980, the U.S. Postal Service issued a postage stamp in his honor.

Chapter 6

Benjamin Montgomery was born into slavery in 1819 in Loudon County, Virginia. He died 1877. He was sold to Joseph E. Davis, a Mississippi planter. Davis was the older brother of Jefferson Davis who would later serve as the President of the Confederate States of America. After a period time, according to (Chamberlain, 2019) Davis could see great talent within Montgomery and assigned to him the responsibility of running his general store on the Davis Bend plantation. Montgomery, who by this time had learned to read and write (he was taught by the Davis children), excelled at running the store and served both white customers and slaves who could trade poultry and other items in return for dry goods. Impressed with his knowledge and abilities to run the store, Davis placed Montgomery in charge of overseeing the entirety of his purchasing and shipping operations on the plantation.

In addition to being able to read and write, Montgomery also learned a number of other difficult tasks, including land surveying, techniques for flood control and the drafting of architectural plans. He was a skilled mechanic and inventor.

Montgomery decided to created a propellor that could cut into the water at different angles, thus allowing the boat to navigate more easily though shallow water. Joseph Davis attempted to patent the device but the patent was denied on June 10, 1858, on the basis that Ben, as a slave, was not a citizen of the United States, and thus could not apply for a patent in his name. Later, both Joseph and Jefferson Davis attempted to patent the device in their names but were denied because they were not the "true inventor." Ironically, when Jefferson Davis later assumed the Presidency of the Confederacy, he signed into law the legislation that would allow a slaves to receive patent protection for their inventions. On June 28, 1864, Montgomery, no longer a slave, filed a patent application for his devise, but the patent office again rejected his application.

Upon the end of the Civil War, Joseph Davis sold his plantation as well as other properties to Montgomery, along with his son Isaiah. Benjamin and Isaiah decided to pursue a dream of using the property to establish a community of freed slaves. (Chamberlain, 2019) He later on bought some land and along with a number of other former slaves, and founded the town of Mound Bayou, Mississippi in 1887. Isaiah was named the town first mayor soon thereafter.

Developed a Special Steamboat Propeller.

David Crosthwait

Designed Innovative Heating Installations. Mechanical Engineer born May 27,1898- died February- 25, 1976 **David Crosthwait** was born in Nashville, Tennessee and moved to Kansas City, Missouri where he went to high school. According to (Chamberlain, 2019) In 1913 Crosthwait moved to Marshalltown, Iowa where he began working for the Durham Company, designing heating installations and central air conditioning. In 1925 he was named the director of the research department, overseeing a staff of engineers and chemists.

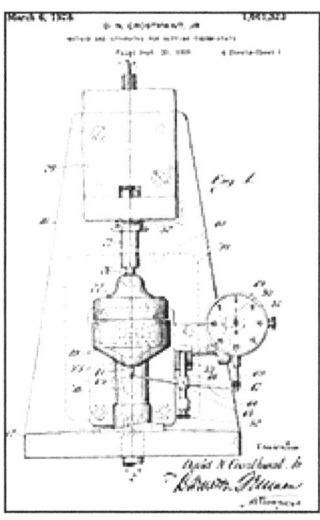

His research concerned heating and ventilating and in the coming years he obtained 39 patents for various devices including heating systems, vacuum pumps, refrigeration methods and processes and temperature regulating devices. His most famous creation was the heating system for New York's famous Radio City Music Hall.

Elijah McCoy

Although the name Elijah McCoy may be unknown to most people in the world, the enormity of his ingenuity and the quality of his inventions have created a level of distinction which bears his name. Elijah McCoy was born in Colchester, Ontario, Canada on May 2, 1844 and died in 1929. His parents were George and Emillia McCoy, former slaves from Kentucky who escaped through the Underground Railroad. George joined the Canadian Army, fighting in the Rebel War and then raised his family as free Canadian citizens on a 160 acre homestead.

According to (Chamberlain, 2019) At an early age, Elijah often taking items apart and putting them back together again. Recognizing his abilities in mechanical, George and Emillia saved enough money to send Elijah to Edinburgh, Scotland, where he could study mechanical engineering. After finishing his studies as a "master mechanic and engineer" he returned to the United States which had just seen the end of the Civil War – and the emergence of the "Emancipation Proclamation."

In an effort to improve efficiency and eliminate the frequent stopping necessary for lubrication of the train, McCoy set out to create a method of automating the task. In 1872 he developed a "lubricating cup" that could automatically drip oil when and where needed. He received a patent for the device later that year. The "lubricating cup" met with enormous success and orders for it came in from railroad companies all over the country. Other inventors attempted to sell their own versions of the device but most companies wanted the authentic device, requesting "the Real McCoy."

McCoy remained interested in continuing to perfect his invention and to create more. He thus sold some percentages of rights to his patent to finance building a workshop. According to (Wikipedia, 2020) He made continued improvements to the "lubricating cup." The patent application described the it as a device which "provides for the continuous flow of oil on the gears and other moving parts of a machine in order to keep it lubricated properly and continuous and thereby do away with the necessity of shutting down the machine sometimes." The device would be adjusted and modified in order to apply it to different types of machinery. Versions of the cup would soon be used in steam engines, naval vessels, oil-drilling rigs, mining equipment, in factories and construction sites.

In 1916 McCoy created the graphite lubricator which allowed new superheater trains and devices to be oiled. In 1920, Elijah established the "Elijah McCoy Manufacturing Company." With his company, he improved and sold the graphite lubricator as well as other inventions which came to him out of necessity. He developed and patented a portable ironing board. When he desired an easier way of watering his lawn, he created and patented the lawn sprinkler. **Elijah McCoy**
McCoy is best known for having invented the automatic oil cup. According to (Chamberlain, 2019) During his life, McCoy invented and sold 57 different kinds of devices and machine parts, including an ironing board and a lawn sprinkler. His first patent was for a lubricator for steam engines (US #129,843), which was issued on July 12th, 1872. In his honor is the "Coined Phrase" "The Real McCoy"
In 1922, Elijah and Mary were involved in an automobile accident and both suffered severe injuries. Mary would die from the injuries and Elijah's health suffered for several years until he died in 1929. McCoy left a legacy of inventions which would benefit for another century and his name would come to symbolize quality workmanship – the Real McCoy!

Ernest Just

Ernest Just was born on August 14, 1883 in Charleston, South Carolina and died June 10, 1858. Ernest was able to earn enough money to attend the Kimball Academy in New Hampshire. He experimented with the reproductive systems and cells of marine animals in the Marine Biological Laboratory in Woods Hole, Massachusetts. According to (Chamberlain, 2019) his research and papers on Marine biology were so well received that in 1915, at age 32, Just was awarded the first Spingarn Medal by the National Association for the Advancement of Colored People.

He performed studies on marine animals and their eggs as well as on their cell structures. He believed that in learning about healthy cells and cell structures, man could hope to understand and find cures for cellular irregularities and diseases such as sickle cell anemia and cancer. He also researched parthenogenesis (developing marine eggs without fertilization). He quickly became one of the most respected scientists in his field, but much of that recognition came from abroad as racial bigotry in the United States caused much of his work and his achievements to go unrewarded. Ernest died on June 10, 1858, of cancer, leaving behind a wife, Ethel and three children. He also left behind a world which would eventually recognize him as the most outstanding zoologist of his time. Pioneer in Marine Biology and Zoology.

Semiconductor and Aerospace Innovator.
George Edward Alcorn, Jr.
Born March 22, 1940- His death date is unkonwn

George is a pioneer in the field of semiconductor devices and one of the top inventors in the field of aerospace. (Wikipedia, 2020) He developed a Semiconductor and is a Aerospace Innovator, George was Born March 22, 1940 in Indianapolis, Indiana, George was the son of Arletta and George Alcorn, Sr., an auto mechanic. Both parents promoted the virtu of education to George, Jr. and his younger brother Charles. He was a good athlete in multiple sports and was very educated student. He then enrolled in Nuclear Physics program at Howard University. He completed his Master's work in 1963.

In 1964, Alcorn applied for a research grant from NASA to study the concept of negative ion formation. He was awarded the grant and conducted his research from 1965 to 1967.

He obtained work during the summers of 1962 and 1963 at North American Rockwell, a leading aerospace company. He worked in the company's the space division and was assigned to perform computer analysis on the orbital mechanics and launch trajectories for rockets and missiles. Some of his work involved the Titan and Saturn rockets from the National Aeronautics and Space Administration's (NASA) Apollo space missions and well as the NOVA missile.

Alcorn signed on with Philco-Ford, a division of the Ford Motor Company. Philco-Ford produced a wide array of products, ranging from car radio to television set. It also had an aerospace division which developed satellite tracking systems for NASA's manned space program. Alcorn served as a senior scientist for the aerospace division. He later worked as a senior physicist for Perkin Elmer, a multinational technology corporation and then as an advisory engineer for International Business Machines (IBM). His relationship with IBM proved quite valuable in 1973 when he was selected to teach as an IBM Visiting Professor in Electrical Engineering at Howard University (eventually becoming a full professor). As if his schedule was not already busy enough, he also taught Electrical Engineering at the University of the District of Columbia as a full professor.

In 1978, Alcorn left IBM and joined NASA where he invented an imaging X-ray spectrometer which used thermomigration of aluminum. X-ray spectrometry is used to provide data which can be analyzed for a number of applications, including for obtaining information about remote solar systems and other space objects. He would receive a patent for the device in 1984. As a result of the significance of this work. he was the NASA/GSFC Inventor of the Year (GSFC is an acronym for the Goddard Space Flight Center, NASA's first space flight center established in May of 1959). In 1986 he developed an improved method of fabrication using laser drilling.

Because of his success in his endeavors, NASA placed him in an administrative/management position as the deputy project manager for advanced development of new technologies for use in the International Space Station, Freedom. In 1990 he was named the manager for advanced programs for NASA/GSFC and in 1992 became the head of the Office of Commercial Programs at GSFC, helping to find commercial uses for the new technologies developed at GSFC. (Chamberlain, 2019) Later he ran the GSFC Evolution program which oversaw the development and running of the space station. In 1994, he oversaw a space shuttle experiment which utilized a "Robot Operated Material Processing System" to conduct the manufacturing of material in the microgravity of space.

In 1999, he was awarded the Government Technology Leadership award and two years later was awarded special congressional recognition for his work for aiding business in the Virgin Island in employing technology. Finally in 2005 he was named the Assistant Director for Standard/Excellent – Applied Engineering and Technology Directorate for GSFC.

Over his career, Alcorn created numerous noteworthy inventions and secured more than 25 patents. He is seen as a pioneer in the field of plasma semiconductor devices.

His concept and implementation of "plasma etching" has become a standard in the industry. He also served his community well over the years, involving himself in programs aimed at recruiting minorities and women to NASA as well as programs to encourage inner-city children to focus on science. In 1984, Alcorn was awarded the NASA-EEO medal for his efforts and was honored by Howard University with its Heritage of Greatness award.

George Carruthers
Born October 1, 1939 and his death is unkown
Measuring and Detected Ultraviolet Lights.
He read with particular interest about the space exploits of the Naval Research Laboratory in Washington, DC and upon graduating from Englewood High School in 1957, he enrolled in the University of Illinois.

Carruthers stayed at the University of Illinois for seven years, receiving a Bachelor of Science degree in Aeronautical Engineering in 1961, a Master's degree in Nuclear Engineering in 1962 and a Ph.D. in Aeronautical and Astronomical Engineering in 1964 (his thesis focused on atomic nitrogen recombination). Upon finishing his Ph.D., he immediately accepted a position with the Naval Research Laboratory (NRL) as a Research Physicist in 1964, having received a fellowship in Rocket Astronomy from the National Science Foundation.

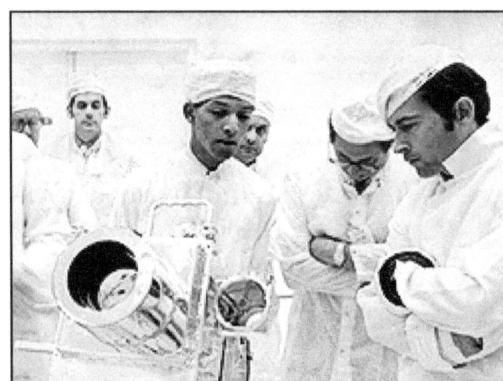 Upon joining the NRL, Carruthers focused his attention on far ultraviolet astronomy, observing the Earth's upper atmosphere and

other astronomical phenomena. In 1966, he became a research assistant at the NRL's E.O. Hulburt Center for Space Research where he began research on ways to create visual images as a means for understanding the physical elements of deep space. He particularly focused on creating a device to analyze and illuminate ultraviolet radiation. His belief was far ultraviolet… is of great importance to the astronomer because it allows the detection and measurements of common elements (hydrogen, oxygen, nitrogen, carbon, and many others) in their cool, unexcited state… This allows more accurate measurements of the compositions of interstellar gas, planetary atmospheres, etc..

According to (Chamberlain, 2019) In 1969, Carruthers received a patent for his invention the "Image Converter for Detecting Electromagnetic Radiation Especially in Short Wave Lengths" which is designed to detect electromagnetic radiation in short wave lengths. His research helped him be the principle inventor of the Far Ultraviolet Camera/Spectrograph which would ultimately be used on the Apollo 16 mission to the moon. Ultraviolet (UV) light is the range of electromagnetic radiation that lies between visible light and X-Rays. The gold-plated camera system was able to record radiation existing in the upper half of the ultraviolet system of the atmosphere. The camera allowed views of stars and celestial bodies and looks into the solar system thousands of miles away, as well as of the earth. A second version of the camera was sent on the 1974 Sky Lab space flight. One of the great uses of the camera was to permit a viewer to visually see the effects of pollution on the atmosphere. The camera also was able, for the first time, to detect hydrogen in space, which gave an indication that plants were not the only source of oxygen for the Earth and led to a renewed debate about the origin of stars.

George Carruthers has continued to offer innovation in the areas of astronomy and physics and has been active in outreach programs seeking to bring science to youth around the country. He has been lauded for his efforts and achievements. He was named Black Engineer of the Year in 1987, awarded the Arthur Fleming Award in 1971, the Exceptional Achievement Scientific Award from NASA in 1972, the Warner Prize in 1973 and was inducted into National Inventors Hall of Fame in 2003. His success is primed to lead to greater achievements by those who follow in his footsteps in the future.

George Crum
Invented the Potato chip

George Crum was born originally named George Speck in 1822 in Saratoga Lake, New York and died 1904. the son of a Huron Native-American mother and an African-American father who worked as a jockey. He intended to make french fries but a guest complained that they were too thick. Annoyed, he prepared another batch and sliced the potatoes extremely thin. After deep frying them in oil he found them very thin and very crisp and after adding salt found that the guests loved them. George began preparing the potatoes this way and they would soon become known as potato chips.

Unfortunately, he never patented the potato chip, nor sought to market them outside of his restaurant. A few years after he retired, however, potato chips were mass marketed by others and would eventually become a six billion dollar a year industry. George Crum died in 1904 at the age of 92.

George Franklin Grant
Created an Improved Golf Tee. Born September 15, died 1846-August 21,1910
He was a dentist, Inventor and professor.

George Murray
Born September 22,1853- and died April 21,1926
Created a Cotton-chopping Device.

Chapter 8

Granville Woods
Born April 23,1856 and died January 30,1910
Telegraph and Railway Device Developer.

Granville Woods was certainly a respected inventor as he was often referred to as the "Black Thomas Edison."

Granville Woods was born on April 23, 1856 in Columbus, Ohio January 30, 1910, he died January 30, 1910. He spent his early years attending school until the age of 10 at which point he began working in a machine shop repairing railroad equipment and machinery. Intrigued by the electricity that powered the machinery, Over the next few years, Woods moved around the country working on railroads and in steel rolling mills. This experience helped to prepare him for a formal education studying engineering.

According to (Chamberlain, 2019) (BlackPast, 2020) Woods obtained a job as an engineer on a British steamship called the Ironsides. Out of frustration for not getting hired because of the color of his skin, even though he was very intelligent. He instead to promote his abilities, (Woods), along with his brother Lyates, formed the Woods Railway Telegraph Company in 1884. The company manufactured and sold telephone, telegraph and electrical equipment. One of the early inventions from the company was an improved steam boiler furnace and this was followed up by an improved telephone transmitter which had superior clarity of sound and could provide for longer range of distance for transmission.

In 1885, Woods patented a apparatus which was a combination of a telephone and a telegraph. The device, which he called "telegraphony," would allow a telegraph station to send voice and telegraph messages over a single wire. The device was so successful that he later sold it to the American Bell Telephone Company. In 1887, Woods developed his most important invention to date – a device he called Synchronous Multiplex Railway Telegraph. A variation of the "induction telegraph," it allowed for messages to be sent from moving trains and railway stations. By allowing dispatchers to know the location of each train, it provided for greater safety and a decrease in railway accidents.

Granville Woods often had difficulties in enjoying his success as other inventors made claims to his devices. Thomas Edison made one of these claims, stating that he had first created a similar telegraph and that he was entitled to the patent for the device. Woods was twice successful in defending himself, proving that there were no other devices upon which he could have depended or relied upon to make his device. After the second defeat, Edison decided that it would be better to work with Granville Woods than against him and thus offered him a position with the Edison Company.

In 1892, Woods used his knowledge of electrical systems in creating a method of supplying electricity to a train without any exposed wires or secondary batteries. Approximately every 12 feet, electricity would be passed to the train as it passed over an iron block. He first demonstrated the device as an amusement apparatus at the

Coney Island amusement park and while it amused patrons, it would be a novel approach towards making safer travel for trains.

Many of Woods inventions attempted to increase efficiency and safety railroad cars, Woods developed the concept of a third rail which would allow a train to receive more electricity while also encountering less friction. This concept is still used on subway train platforms in major cities in the United States.

Over the course of his life time Granville Woods would obtain more than 50 patents for inventions including an automatic brake and an egg incubator and for improvements to other inventions such as safety circuits, telegraph, telephone, and phonograph. When he died on January 30, 1910 in New York City he had become an admired and well respected inventor, having sold a number of his devices to such giants as Westinghouse, General Electric and American Engineering – more importantly the world knew him as the Black Thomas Edison.

JACK JOHNSON Boxer (Born March 31,1878- Died June 10,1946)
Boxing Champion Invented a Wrench.

James Forten

Was a free black man in Philadelphia, Pa. He purchased some slaves freedom. Born 1766- Died in1842. He revolutionized the Sail-making Business which made it easier for sailors to sail.

James West
Born 10, 1931- present
Created the Affordable Electret Microphone.

Joseph Dickinson
Born in Canada 1955-present
Developed Improved and Innovative Reed Organs. Piano reverse and forward mechanism.

Joseph Lee

Born July 19,1849-June 11,1908 Created a Bread Making Machine.

Chapter 9

Lewis Temple
Born Oct 1,1800-May 18,1854
Created the "Temple's Iron" Harpoon.

Lloyd Hall

Born June 20,1894-Diedjanuary 2, 1971
Revolutionized the World of Food Preservation.

Lonnie Johnson
Invented the "Super Soaker" toy.

You don't have to be a rocket scientist to come up with a great idea, but it certainly doesn't hurt. For Lonnie Johnson, a lifetime of achievement and success at various levels on government and private sector projects could not prepare him for the success he would ultimately achieve – by building a better squirt-gun.

Lonnie Johnson was born on October 6, 1949 in Mobile, Alabama. His father worked as a civilian driver at Brookley Air Force Base, and his mother was a homemaker who worked part time as a nurse's aide. His father taught Robert and his brothers how to repair various household items, prompting the boys to create their own toys. The boys once made a go-kart out of household items and a lawn mower motor. He had of talents and also attended Williamson High School and in 1968, as a senior, took part in a national science competition sponsored by the University of Alabama. There he displayed a remote controlled robot named "Linex" which he built from scraps found at a junkyard and parts of his brothers' walkie-talkie and his sisters' reel-to-reel tape recorder. He has a Bachelor of Science degree in Mechanical Engineering. He continued on at Tuskeegee and received a Master's Degree in Nuclear Engineering in 1975.

After he graduated, he took a position at the Savannah River National Laboratory, conducting thermal analysis on plutonium fuel spheres. He later served as a research engineer, developing cooling systems at the Oak Ridge National Laboratory in Oak Ridge, Tennessee. Lonnie Johnson then joined the Air Force and was assigned to the Air Force Weapons Laboratory in Albuquerque, New Mexico where he served as the Acting Chief of the Space Nuclear Power Safety Section. In 1973, he left the Air Force and took over as Senior Systems Engineer at NASA's Jet Propulsion Laboratory in Pasadena, California. He worked on the Galileo Mission to Jupiter, but returned in 1982 to his military career. He worked at the Strategic Air Command (SAC) facility in Bellevue, Nebraska and then moved to the SAC Test and Evaluation Squadron at Edwards Air Force Base in Edwards, California where he worked on the Stealth Bomber. He also worked as Acting Chief at the Space Nuclear Power Safety Section of the Air Force Weapon Laboratory at Kirkland Air Force Base in New Mexico. A Captain, he was awarded the Air Force Achievement Medal and the Air Force Commendation Medal.

Earlier, around 1982, he was working on developing a heat pump that would work by circulating water rather than expensive and environmentally unfriendly freon. In his basement at home, he took some tubing with a specially devised nozzle on the other end and connected it to a bathroom sink. When he turned on the faucet, a stream of water shot out of the nozzle across the room with such force that the air currents caused the curtain to move. His first thought was "this would make a great water gun."

Johnson set out to develop a pressurized water gun that was safe enough for children to play with. Water guns at the time very unsophisticated and cheaply made, able to shoot streams of water about eight feet. His original super soaker he called the Power Drencher, he and his partner Bruce D'Andrade began trying to market it while trying to secure a patent for it. The Super Soaker, that Johnson invented was later on introduced to Al Davis, an executive with Larimi Corp. at a New York City Toy Fair. Two weeks later Johnson was in Larimi's headquarters in Philadelphia.

Johnson continued inventing and would eventually hold more than 80 patents. For his contributions to science (and in light of his great success with the Super Soaker) Johnson was inducted into the Inventor Hall of Fame in 2000. His company has continued to innovate, creating improved radon detectors, heat pumps and lithium battery products as well as new toy concepts.

Mark Dean

Born March 2, 1957-present

Created innovative Computer Devices and a computer chip.

He was the co-creator and of the IBM computers. Then came VP of IBM company He also invented the colored pc monitors.

Marjorie Joyner

Born October 24,1896- Died December 27,1994
Invented the "Permanent Waving Machine."

Meredith Charles Gourdine
Born September 26,1929-November-20,1998
Created an Allergen-filtration Device.

Miriam Benjamin

Born September 16, 1861- Died 1947
Invented a "Gong and Signal" chair for hotels.

Chapter 10

Norbert Rillieux

Created Sugar-refining Evaporation Pan System. Born March 17, 1806- Died October 8, 1894

Otis Boykin

Born August 29,1920- Died March 13,1982
Invented Electrical Resistors used in computers and missile guidance and Pacemakers. He was a great Inventor and Engineer.

Patricia Bath

Born November 4, 1942- Died May 30, 2019
Created a Laser Surgical Device. She Invented laserphaco probe for cataract treatment in 1986

Philip Emeagwali

Born August 23,1954-present

Created Processes to Innovate Internet Apps and microprocessors. He also Invented the fastest computer. He helped led to the development of the internet.

Richard Bowie Spikes

Born October 2,1878- Died January 22,1963

Invented an Automatic Gear Shift and automobile directional signals and beer tap.

Robert Pelham

Born January 4,1859- Died May 12,1943
Invented Tallying Machine for the Census. He was a Journalist and Inventor.

Sarah Elisabeth Goode
Invented the Folding Cabinet Bed.
Born 1855- Died April 8, 1908

Shelby Davidson
 Was born1868- and then died in1931.
Created Adding Machine Add-ons. Was a lawyer

Chapter 11

Thomas Elkins
Born 1818- Died August 10, 1900
Created a Refrigerated Apparatus and mirror and blood banks. According to (Chamberlain, 2019) He was also Surgeon, Dentist Inventor. Medical Examiner for 54 and 55 Massachusetts infantries in the civil war.

Thomas Mensah
Born in 1950 from Kumasi Ghana.
Pioneered the Field of Fiber Optics and nonotechnology. The use of light pulses to transmit data through cables from one place to another, which is known as fiber optics communication. He was Chemical Engineer that holds 14 patents.

Valerie L. Thomas was Born February 8,1943-present Developed a 3-D Optical Illusion Device. She was responsible for developing the digital media formats image process system.

Born March 4, 1877- July 27, 1963

Garrett Morgan is one of those rare people who are able to come up with an extraordinary inventions which have a tremendous impact on society – and then follows that up with even more!

Garrett Morgan was born on March 4, 1877 in Paris, Kentucky the seventh. AACSB accredited online MBA or masters in education. In Cleveland, he learned the inner workings of the sewing machine and in 1907 opened his own sewing machine store, selling new machines and repairing old ones.

According to (Chamberlain, 2019) In 1909, Morgan opened a tailoring shop, selling coats, suits and dresses. While working in this shop he came upon a discover which brought about his first invention. He noticed that the needle of a sewing machine moved so fast that its friction often scorched the thread of the woolen materials. He thus set out to develop a liquid that would provide a useful polish to the needle, reducing friction. In 1912, Morgan developed another invention, much different from his hair straightener. Morgan called it a Safety Hood and patented it as a Breathing Device, but the world came to know it as a Gas Mask.

The Safety Hood consisted of a hood worn over the head of a person from which emanated a tube which reached near the ground and allowed in clean air. The bottom of the tube was lined with a sponge type material that would help to filter the incoming air. Another tube existed which allowed the user to exhale air out of the device. Morgan intended the device to be used "to provide a portable attachment which will enable a fireman to enter a house filled with thick suffocating gases and smoke and to breathe freely for some time therein, and thereby enable him to perform his duties of saving life and valuables without danger to himself from suffocation. The device is also efficient and useful for protection to engineers, chemists and working men who are obliged to breathe noxious fumes or dust derived from the materials in which they are obliged to work."

The National Safety Device Company, with Morgan as its General Manager was set up to manufacture and sell the device and it was demonstrated at various exhibitions across the country. At the Second International Exposition of Safety and Sanitation, the device won first prize and Morgan was award a gold medal.

Although he could have relied on the income his Gas Masks generated, Morgan felt to develop a automatically directing traffic without the need of a policeman or worker present. He patented an automatic traffic signal which he said could be "operated for directing the flow of traffic" and providing a clear and unambiguous "visible indicator."

Satisfied with his efforts, Morgan sold the rights to his device to the General Electric Company for the astounding sum and it became the standard across the country. Today's modern traffic lights are based upon Morgan's original design. He died later on July 27, 1963 and because of his contribution, the world is certainly a much safer place.

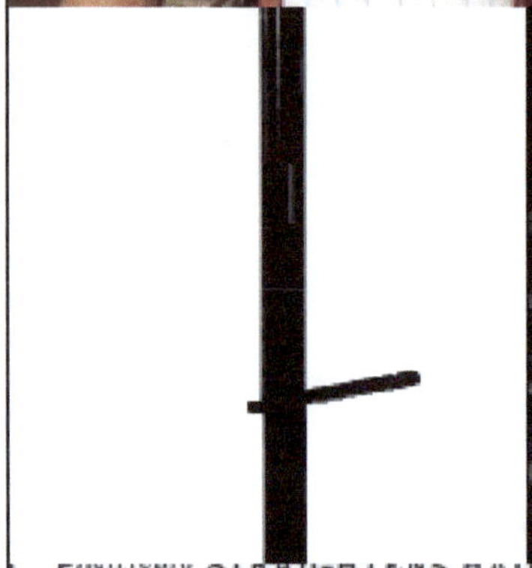

Positively STRAIGHTENS HAIR in 15 Minutes

G. A. MORGAN'S SPECIAL HEAVY HAIR PRESSING COMB

"THE BEST THERE IS—CHEAPER THAN THE CHEAPEST."

G. A. MORGAN'S HAIR PRODUCTS

"THE ONLY COMPLETE LINE OF HAIR PREPARATIONS IN THE WORLD"

HAIR REFINER CREAM—Positively straightens hair in 15 minutes $1.00
HAIR REFINER SHAMPOO SOAP—Necessary for treatment with Hair Refiner Cream, and a beautiful shampoo .... .25
ITALIAN HAIR OIL—Beautifies the hair and disappears .... .35
HAIR GROWER—Promotes an excellent growth of good looking hair .... .50
HAIR PRESSING GLOSS—Makes hair soft, straight and glossy .. .50
DANDRUFF AND TETTER OINTMENT—Relieves worst cases of Dandruff and Tetter 1.00
HAIR PRESSING NIGHT CAP—Presses and trains the hair

Chapter 12

Fred Mckinley Jones
Born May 17,1893 Died February 21 1961
He definitely is certainly one of the most important Black inventors ever based on the his inventions he formulated as well as their diversity. Fred Jones was born on May 17, 1893 in Covington, Kentucky. Died February 21 1961, He Eventually became interested in automobiles.

He help design and repair automobiles that were brought in by customers but also served as a studio for building racing cars. According to (Chamberlain, 2019) After a few years of building these cars, Fred desired to drive them and soon became one of the most well known racers in the Great Lakes region. After brief stints working aboard a steamship and a hotel, Jones moved to Hallock, Minnesota began designing and building racecars which he drove them at local tracks and at county fairs.

He convert scrap metal into the parts necessary to deliver a soundtrack to the video, he also devised ways to stabilize and improve the picture quality. Fred began improving on many of the existing devices the company sold. Many of his improvements were so significant, representatives from A.T. & T and RCA sat down to talk with Fred and were amazed at the depth of his knowledge on intricate details, particularly in light of his limited educational background. Around this time, Fred came up with a new idea – an automatic ticket-dispensing machine to be used at movie theaters. Fred applied for and received a patent for this device in June of 1939 and the patent rights were eventually sold to RCA.

Jones was a Pioneer for the United States Army Infantry and served in France during World War I. While serving, Jones recruited German prisoners of war and rewired his camp for electricity, telephone and telegraph service. When navigation through the snow proved difficult, Fred attached skis to the undercarriage of an old airplane body and attached an airplane propeller to a motor and soon whisked around town at high speeds in his new snow machine. When people were complaining about x-rays. Jones told them people to come into his office for x-ray exams, Jones created a portable x-ray machine that could be taken to the patient. Unfortunately, like many of his early inventions, Jones never thought to apply for a patent for machine and watched helplessly as other men made fortunes off of their versions of the device. Undaunted, Jones set out for other projects, including a radio transmitter, personal radio sets and eventually motion picture devices.

At some point, Joe Numero was presented with the task of developing a device which would allow large trucks to transport perishable products without them spoiling. Jones set to work and developed a cooling process that could refrigerate the interior of the tractor-trailer. In 1939 Fred and Joe Numero received a patent for the vehicle air-conditioning device which would later be called a Thermo King.

In addition to installing the Thermo King refrigeration units in trucks and tractor-trailers, Jones modified the original design so they could be outfitted for trains, boats and ships.

During World War II, the Department of Defense found a great need portable refrigeration units for distributing food and blood plasma to troops in the field. The Department called upon Thermo King for a solution. Fred modified his device and soon had developed a prototype which would eventually allow airplanes to parachute these units down behind enemy lines to the waiting troops.

For the next 20 years, Fred Jones continued make improvements on existing devices and devised new inventions when necessary to aid the public. Jones died on February 21, 1961 and was posthumously awarded the National Medal of Technology, one of the greatest honors an inventor could receive. Jones was the first Black inventor to ever receive such an honor.

Chapter 13

Thomas Alva Edison is the Inventor of the light bulb and other inventions (Born February 11, 1847 – and then died October 18, 1931) and also was
an American inventor and businessman. He developed many devices that greatly influenced life around the world, including the phonograph, the motion picture camera, and the long-lasting, practical electric light bulb. Dubbed "The Wizard of Menlo Park", he was one of the first inventors to apply the principles of mass production and large-scale teamwork to the process of invention, and because of that, he is often credited with the creation of the first industrial research laboratory.

According to (Chamberlain, 2019) Edison was a prolific inventor, holding 1,093 US patents in his name, as well as many patents in the United Kingdom, France, and Germany. More significant than the number of Edison's patents was the widespread impact of his inventions: electric light and power utilities, sound recording, and motion pictures all established major new industries world-wide. Edison's inventions contributed to mass communication and, in particular, telecommunications. These included a stock ticker, a mechanical vote recorder, a battery for an electric car, electrical power, recorded music and motion pictures.

Charles E. Henderson Jr. Portable Air Conditioning System patented November 10, 2015

Born August 23,1981- Present

Charles Invented the Portable Air Conditioning System in Nov 10, 2015 that can be used indoor and outdoors. He is a music marketing manager and promotion manager, book author, inventor of "The Portable Air Conditioning System", Charles also is the original developer of "Sports crush game app" in 2019 Charles Edward Henderson Jr has been an engineer and very handy for well over 15yrs. Charles was very Athletic in Baseball and Football and Track. He got scouted for the MLB but didn't turn professional, but he ran track in College for Slippery Rock University because he sacrificed his career to be more around his kids and wife. He is happily married to his wife Ashley Henderson and they have a few kids. He been very athletic and family oriented and loves God and Jesus and his family an into the bible alot and was Born in Pittsburgh Pa 1981- present

William Henry "Bill" Gates III is the **Inventor of Microsoft** (born October 28, 1955-present) He is an American businessman and magnate, philanthropist, investor, computer programmer, and inventor. In 1975, Gates co-founded Microsoft, that became the world's largest PC software company, with Paul Allen. During his career at Microsoft, Gates held the positions of chairman, CEO and chief software architect, and was the largest individual shareholder until May 2014. Gates has authored and co-authored several books as well. He has a foundation to help others.

 **The First Automobile was made invented by Karl Friedrich Benz** (German:) November 25, 1844 – April 4, 1929) who is a German engine designer and engineer, generally regarded as the inventor of the first automobile powered by an internal combustion engine, and together with Bertha Benz, pioneering founder of the automobile manufacturer Mercedes-Benz. According to (BlackPast, 2020) Other German contemporaries, Gottlieb Daimler and Wilhelm Maybach working as partners, also worked on similar types of inventions, without knowledge of the work of the other, but Benz received a patent for his work first, and, subsequently patented all the processes that made the internal combustion engine feasible for use in an automobile. In 1879, his first engine patent was granted to him, and in 1886, Benz was granted a patent for his first automobile.

Inventor **Charles Babbage** is the First Mechanical Computer, known as the father of the Computer.

Charles Babbage   December 1791 –18 October 1871) from London United Kingdom. He was an English polymath. He was a mathematician, philosopher, inventor and mechanical engineer, Babbage is best remembered for originating the concept of a programmable computer.

According to (BlackPast, 2020) He was Considered a "father of the computer" and babbage is credited with inventing the first mechanical computer that led to more complex designs. His varied work in other fields has led him to be described as "pre-eminent" among the many polymaths of his century.

Parts of Babbage's uncompleted mechanisms are on display in the London Science Museum. In 1991, a perfectly functioning difference engine was constructed from Babbage's original plans.

Dr. Gladys West is a very phenomenal well educated black woman who Invented the Global Positioning System GPS, who was honored by the U.S. Air Force at the Pentagon. Born 1930- present

John Parker
1827- January 30,1900
Created a Screw for a Tobacco Press.
By the 1880s Parker became an inventor. In 1884, 1885, and 1890, the U.S. Patent Office approved his patents for an improved tobacco press, a portable tobacco press, and soil pulverizer, respectively. According to (Chamberlain, 2019) He also expanded into the flour milling business and his product was displayed at the New Orleans Exposition in 1884. John Parker, inventor and businessman, was also a prominent Underground Railroad conductor before the Civil War. He was reputedly responsible for the rescue of nearly 1,000 enslaved people between 1845 and 1865. Parker repeatedly crossed the Ohio River from his home in Ripley, Ohio, often going as far as 20 miles on foot into Kentucky to rescue fugitive slaves and bring them to freedom.

Albert Richardson
Born May 19,1880- Feb 3,1964
Invented a Wooden Butter Churn.

Benjamin Bradley
1830-1897
Created a Steam Engine for Warships.

Benjamin Thornton

1819-1877
Created an Answering Machine Device.

Charles Brooks
1865-November 19, 1956
Developed the Street-sweeper.

Daniel McCree
Created a Portable Home Fire Escape. He was a boot maker that developed trimming soles of boots.
Born a slave around 1835- death is unknown.

David Fisher
Born April 13,1929- Died (Chamberlain, 2019)January 10,2018
Creating the "Joiner's Clamp.

Edward Lewis
Born May 20,1918- Died July 21,2004
Invented a Spring Gun.

Henry Blair
Born 1807- Died 1860
Invented a Seed Planter and Corn Harvester. Helped out agriculture a lot.

Henry Brown
Created a Strongbox for Personal Valuables. 1849-1897

Henry Faulkner
Created a Ventilated Shoe to Minimize Blisters brining heathy feet.
Born January 9, 1924 Death date-December -3,1981

John Love
Invented Early Pencil Sharpener. 1889-1931

Joseph Hawkins
Born November 14, 1781- Died 1823
Improved Gridiron Metal Oven Racks.

Lloyd Ray
Created the Dust Pan. And a mechanical pencil sharpener on the wall. Born 1860- He Died April 21, 1940

Matthew Cherry
Created a Tricycle in 1888 and a Streetcar Fender. Feb 5, 1834-? No information about his date of his death.

Philip Downing
Created a Street Postal Mailbox.   (1857-1934)

Sarah Boone
(1832–1904) was an American inventor who on April 26, 1892, obtained United States patent number 473,563[1] for her improvements to the ironing board.
Created a Version of the Ironing Board.

Thomas Jennings
(1791 – February 12, 1856)
Developed a "Dry-cleaning" process. Received his patent in 1821.

Thomas Stewart
(1846-1935)
Invented a New Version of the Mop June 11,1893

William Purvis

(1838-1914)

Invented a Fountain Pen. In addition to his fountain pen, Purvis, a resident of Philadelphia, Pennsylvania, also successfully patented numbers of other inventions. Between 1884 and 1897 he patented bag machines, a bag fastener, a hand stamp, an electric railway device, an electric railway switch and a magnetic car balancing device. He also is believed to have invented, and yet not patented several other devices such as the edge cutter found on aluminum foil, cling wrap and wax paper boxes.

Willis Johnson

(1857-1923)

On February 5, 1884, Willis Johnson patented a device made up of a handle attached to a series of spring-like whisk wires used to help mix ingredients. Prior to his eggbeater, all mixing of ingredients was done by hand and was quite labor-intensive and time-consuming.

W.H. Richardson (Inventor)
Born on December 5, 1808 Died on December 14, 1870   baby buggy

L.R. Johnson (Inventor) bicycle frame
Born On November 1, 1803 Died on March 15, 1853

A.P. Ashbourne   biscuit cutter
Year Born 1820   Year of Death 1915

Charles Drew   blood plasma bag
June 3, 1904   Died on April 1, 1950

T. Elkins   chamber commode
Year of Birth 1818   Died on August 10, 1900

G.T. Sampson   clothes dryer
Born on April 22, 1934   Died On June 4, 2015

S.R. Scratton   curtain rod in 1892
Born in February 1841   Died on October 14, 1908

O. Dorsey (Inventor)
Born on September 19, 1862 Died on September 15, 1913   December 10, 1878 door knob and door step

Lawrence P. Ray (Inventor)
Born on January 7, 1860 and Died on April 25, 1940   Dust-Pan August 3, 1897

Alexander Miles (Inventor)
Born on May 18, 1838 and Died On May 7, 1918   October 11, 1867 Elevator Door

James P. Johnson (Inventor)
Born on February 1, 1894 in New Jersey Died on November 17, 1955 in New York November 2, 1880 Eye Protector

J.W. Winters invented the fire escape ladder on May 7, 1878
Born on August 29, 1860 in Virginia   Died on November 29, 1916

T.J Marshall Invented the fire extinguisher on October 26, 1872
Marshall was Born on April 5, 1828 in West Virginia   then later died on October 8, 1901

L. C. Bailey   folding bed July 18, 1899
Year of Bailey Birth 1825   Died on September 1, 1918

Alice H. Parker   Invented the furnace
She was born 1895- and died 1920 She made a big impact on the heating Industry.

Robert F. Flemming, Jr.   guitar   March 3, 1886
Month and Year of Birth July 1839 in Baltimore Maryland and Died on February 23, 1919 Melrose Ma

Lyda O. Newman   hair brush
Year of Birth 1885 Ohio and Death is Unknown in New York

Walter B. Purvis   hand stamp   February 27, 1883
Born on August 12, 1838 Died on August 10, 1914

Augustus J. Ricks   horse shoe   March 30, 1885
Year of Birth 1843   Died on 1906

A.L. Cralle (Inventor) ice cream scooper February 2, 1897
Born on September 4, 1866 in Virginia Died on May 3, 1920 in Pittsburgh, Pennsylvania

Norbet Rillieux (Inventor) improv. sugar making   December 10, 1846
Born on March 17, 1806 in New Orleans Louisiana and Died on October 8, 1894 in Paris, France

Albert Richardson (Inventor)   insect-destroyer gun and Casket Lowering Device February 28, 1899
Born on May 19, 1880 in London England and Died on February 3, 1964

F. J. Loudin (Inventor) invented the key Chain   January 9, 1894
Born on January 1, 1836 in Charlestown Ohio and Died on November 3, 1904 in Ravenna Ohio

Michael C. Harvey (Inventor)   lantern August 19, 1884
Year of Birth 1854 in Saint Louis Missouri   Died Unknown

John Albert Burr (Inventor) invented the lawn mower   May 19, 1889
Year of Birth 1848 in Maryland and Died in 1926

John Thomas White (Inventor)   lemon squeezer December 8, 1893
Born on September 3, 1828 in Warwick County, Indiana and Died on March 3, 1996

W. A. Martin (Inventor)   lock   in   July 21, 1889
Born on June 1, 1878 in Chattanooga Tennessee and Died on August 5, 1955

Phillip B. Downing (Inventor)   mailbox on October 27, 1891
Year of Birth 1857   Died in the year of 1934

Joseph Hunger Dickinson (Inventor)   Piano Player on January 8, 1819
Born on June 22, 1855 in Chatham Ontario Canada and Died Unknown

John Standard (Inventor)   Refrigerator and Oil Stove on June 14, 1891
Born on June 15, 1868 and Died in the year of 1900

W. D. Davis (Inventor)   Riding Saddles on   October 6, 1895
Born on May 29, 1853 in Newport Herkimer County in New York and Died in the year of 1924 in Leavenworth County Kansas.

Charles Orren Bailiff (Inventor) of the Shampoo Headrest on October 11, 1898
Year of Birth 1866   Year of Death 1952

Edmond Berger (Inventor) Invented Spark Plug on February 2, 1839
Year of Birth 1898   Year of Death 1968

Rene Laennec (Inventor) invented the Stethoscope
Born on February 17, 1781 and Died on August 13, 1826

T. A. Carrington (Inventor) the Stove on July 25, 1876
Born on November 17, 1842 in London England and Died on October 9, 1918

Burridge & Marshman   Typewriter April 7, 1885
                         Musical devices
Born December19,1946   and died 1930

Miriam E. Benjamin
Ms. Benjamin was the second black woman to receive a patent. She received a patent for an invention she called 'a Gong and Signal Chair for Hotels.'
Born September 16,1861- died 1947

Henry Blair
Born 1807- and died1860
Henry Blair was the second black inventor issued a patent by the United States Patent Office.

Bessie Blount
November24,1914 and died December 30,2009
Blount invented a device to help disabled persons eat.

Henry Brown
Born 1815 in Virginia and died on June 15,1897
Brown patented a 'receptacle for storing and preserving papers' on November 2, 1886, which developed into what is now known as the bank safety deposit box

Emmett W. Chappelle
Born October 25,1925 and died October 14,2019
Chappelle was a biochemist, photobiologist, astrochemist and inventor.

John B. Christian
Born 1927- and died September 26, 2018
John B. Christian invented and patented new lubricants that are used in high flying aircraft and NASA space missions.
David Crosthwait
Crosthwait holds 39 patents for heating systems and temperature regulating devices. He is most well-known for creating the heating system for New York City's famous Radio City Music Hall.

Dr. Charles Richard Drew
Born June 3, 1904 and died April 1, 1950
Drew was the first person to develop the blood bank which are in different places all over the world.

Philip Emeagwali
Philip was born August 23,1954- present
In 1989, Emeagwali won the Gordon Bell Prize, that was considered the equivalent of the Nobel Prize, for developing the fastest supercomputer software in the world.

Sarah E. Goode
Born1885- and died in Chicago 1905
Sarah Goode was the first African-American woman to receive a patent (US #322,177), for which she invented a type of cabinet bed.

Meredith C. Gourdine
Gourdine was the inventor of electrogasdynamics systems.

Lloyd Augustus Hall
Born on June 20,1894 and died January 2,1971
Lloyd Hall is responsible for the meat cutting products, seasonings, emulsions, bakery products, antioxidants, protein hydrolysates and many other products that keep our food fresh and flavorable.

Thomas L. Jennings
Born 1791- and died February 12, 1856
Thomas L. Jennings was the first African American to receive a patent (US patent3306x), which was issued on March 3rd, 1821.

Percy Lavon Julian
Percy was born April 11,1899 and died April 19,1975
Julian synthesized the medicines physostigmine for glaucoma and cortisone that is used for rheumatoid arthritis. He invented fire-extinguishing equipment.

References/Citations

www.blackpast.org (BlackPast, 2020)

www.blackinventor.com (Chamberlain, 2019)

www.wikipedia.com (Wikipedia, 2020)